STEM Superstars

Margaret Hamilton

by Rachel Castro

Norwood House Press

Cover: Margaret Hamilton helped send astronauts to the moon.

Norwood House Press

For information regarding Norwood House Press, please visit our website at: www.norwoodhousepress.com or call 866-565-2900.

PHOTO CREDITS: Cover: © Carol Francavilla/AP Images; © age fotostock/Alamy, 15; © claudiodivizia/iStockphoto, 10; © jamie.sue.photography/Shutterstock Images, 5; © JSC/NASA, 20; © Mark Williamson/Science Source, 16; © NASA Image Collection/ Alamy, 9; © Pictures Now/Alamy, 6; © Science History Images/Alamy, 19; © Wangkun Jia/iStockphoto, 13

Hardcover ISBN: 978-1-68450-835-8
Paperback ISBN: 978-1-68404-633-1

Library of Congress Cataloging-in-Publication Data
Names: Castro, Rachel, author.
Title: Margaret Hamilton : NASA programmer / by Rachel Castro.
Description: Chicago, IL : Norwood House Press, [2021] | Series: STEM superstars | Includes index. | Audience: Grades K-1
Identifiers: LCCN 2019058391 (print) | LCCN 2019058392 (ebook) | ISBN 9781684508358 (hardback) | ISBN 9781684046331 (paperback) | ISBN 9781684046379 (ebook)
Subjects: LCSH: Hamilton, Margaret Heafield, 1936---Juvenile literature. | Computer software developers--United States--Biography--Juvenile literature. | Computer programmers--United States--Biography--Juvenile literature. | Women scientists--United States--Biography--Juvenile literature. | Scientists--United States--Biography--Juvenile literature. | Project Apollo (U.S.)--History--Juvenile literature. | Moon--Juvenile literature.
Classification: LCC QA76.2.H36 C37 2021 (print) | LCC QA76.2.H36 (ebook) | DDC 005.1092 [B]--dc23
LC record available at https://lccn.loc.gov/2019058391
LC ebook record available at https://lccn.loc.gov/2019058392

328N—072020
Manufactured in the United States of America in North Mankato, Minnesota.

Table of Contents

Chapter 1

Studies

Margaret Hamilton was born in Indiana in 1936. She liked car rides with her father. They talked about the meaning of life. She loved math too.

Margaret Hamilton grew up in the countryside in Indiana.

Hamilton went to college. She studied math and philosophy. Then she taught high school. She taught French and math.

Hamilton went to Earlham College in Indiana.

Hamilton got a new job in 1959. She worked with computers. Her computer **predicted** the weather.

Hamilton would keep working with computers for many years.

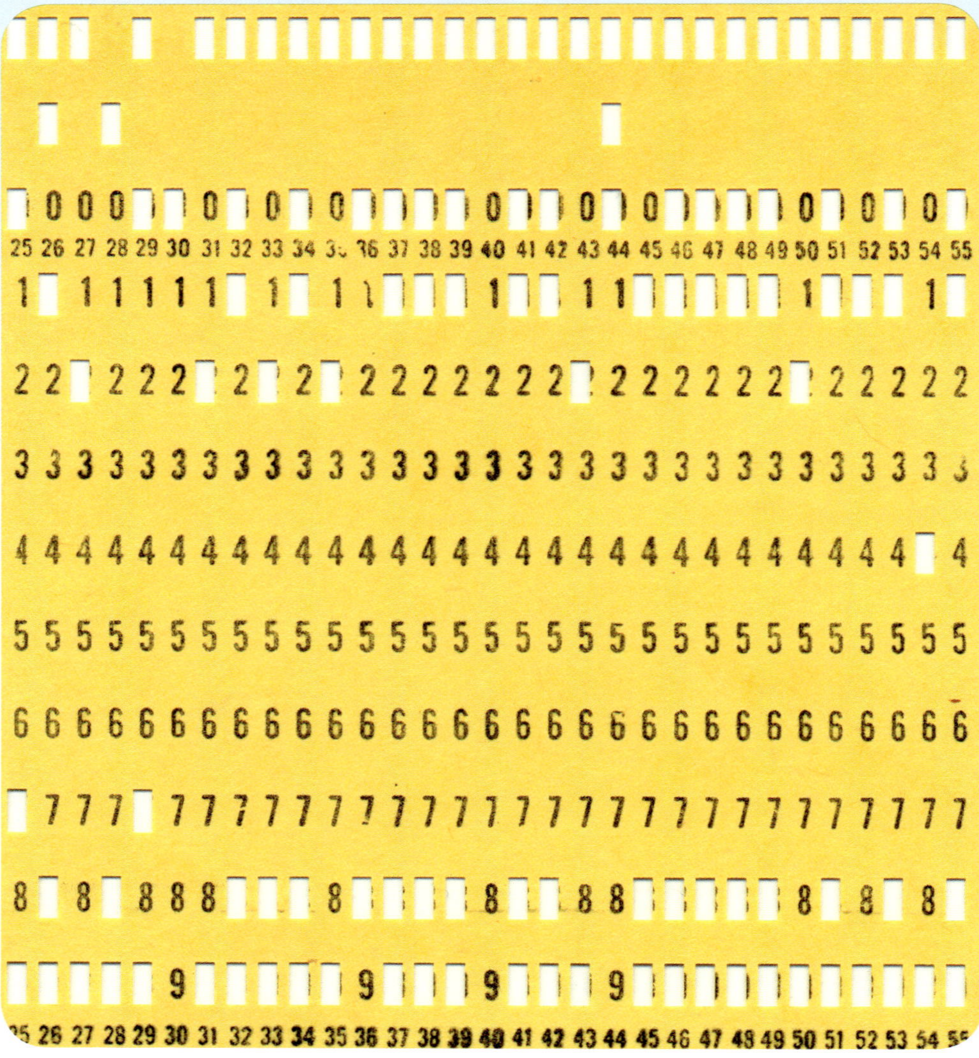

One way people wrote code was by punching holes in cards.

Did You Know?

There were no code classes in the 1960s. Hamilton taught herself.

She learned to write **code**. Computers could read this language.

Chapter 2

Coding for NASA

Hamilton planned to go back to school. She would study more math. But she heard about a new coding job. She changed her plans.

Hamilton's new job was at the Massachusetts Institute of Technology (MIT) in Cambridge, Massachusetts.

The job was for the **National Aeronautics and Space Administration (NASA)**. NASA was sending people to the moon. Its **spacecraft** needed computers. The computers told the spacecraft what to do. They helped the **astronauts**.

Hamilton sits in the Apollo spacecraft. Soon NASA would send it to the moon.

Computers at NASA read the code based on how the magnets were threaded on the wires.

Did You Know?
Hamilton's computer programs were stored on hundreds of tiny magnets.

Hamilton joined the space program. She was the first coder hired in her department. Soon she led her own team. She would help people land on the moon.

Chapter 3

Moon Landing

Apollo 11 launched in 1969. Three minutes before the moon landing, the computer broke. Hamilton knew this might happen. Her code helped the computer recover.

Printed on paper, Hamilton's code filled many books.

MTOA
CRIA
MTIA
MT2A
CRIA
CPIA
CPIA
TYIA
CRIA
LPIA
MT2A

The first people walked on the moon on July 20, 1969.

Did You Know?
Hamilton founded several companies. They make computer programs.

The astronauts landed. Hamilton and her team kept them safe. Her work let astronauts take six trips to the moon. Today all coders use some form of her methods.

★ Career Connections ★

1. Learn how to code by playing a game. There are many free online resources for kids to learn coding.

2. If your school has a coding club, join it. If not, talk to teachers about starting one.

3. Some libraries have STEM activities or workshops for kids. Some even have STEM kits (like a weather experiment or robotics kit) you can check out and work on at home.

4. Hamilton studied philosophy as well as math. Philosophy helps people think about big, important questions about life. People get new ideas for solving problems. You can start learning philosophy by asking questions like *why* and *what if* every day. Ask why something works the way it does or what happens if you try something out.

Glossary

astronauts (AS-truh-nawts): People who are trained to travel into space.

code (KODE): Instructions that tell a computer, phone, or app what to do.

National Aeronautics and Space Administration (NASA) (NASS-uh): The US government group that is in charge of exploring space.

predicted (pri-DIKT-ed): Said or guessed that an action or event would happen in the future.

spacecraft (SPAYS-kraft): A machine designed to travel in outer space.

For More Information

Books

Dean Robbins. *Margaret and the Moon: How Margaret Hamilton Saved the First Lunar Landing*. New York: Alfred A. Knopf, 2017. Read this book to learn more about Margaret Hamilton.

Kiki Prottsman. *My First Coding Book*. New York: DK Publishing, 2017. Learn the concepts behind coding without using a computer.

Websites

NASA Kids' Club

(https://www.nasa.gov/kidsclub/index.html) Play games and learn about NASA at this website.

Wonderopolis: Can You Code?

(https://www.wonderopolis.org/wonder/can-you-code) Learn about why coding is important and find resources to get started.

Index

About the Author

Rachel Castro is a Minneapolis-based writer. She holds degrees in English literature and creative nonfiction. In addition to writing for the educational market, she works for a public library and teaches creative writing.